Collins

easy learning

Maths

Ages 3–5

Peter Clarke

How to use this book

- Find a quiet, comfortable place to work, away from distractions.

- Ask your child what maths they are doing at school and choose an appropriate topic.

- Tackle one topic at a time.

- Help with reading the instructions where necessary and ensure that your child understands what to do.

- Help and encourage your child to check their own answers as they complete each activity.

- Discuss with your child what they have learnt.

- Let your child return to their favourite pages once they have been completed, to play the games and talk about the activities.

- Reward your child with plenty of praise and encouragement.

Special features

- Games: There is a game on each double page, which reinforces the maths topic. Some of the games require a spinner. This is easily made using a pencil, a paperclip and the circle printed on each games page. Place the pencil and paperclip at the centre of the circle and then flick the paperclip to see where it lands.

- Parent's notes: These are divided into 'What you need to know', which explain the key maths idea, and 'Taking it further', which suggest activities and encourage discussion with your child about what they have learnt. The words in bold are key words that you should focus on when talking to your child.

Published by Collins
An imprint of HarperCollinsPublishers Ltd
The News Building
1 London Bridge Street
London
SE1 9GF

Browse the complete Collins catalogue at
www.collins.co.uk

© HarperCollinsPublishers Ltd 2006
This edition © HarperCollinsPublishers Ltd 2015

10 9 8 7 6 5 4 3 2 1

ISBN 978-0-00-815153-9

The author asserts the moral right to be identified as the author of this work.

British Library Cataloguing in Publication Data.

A Catalogue record for this publication is available from the British Library.

Written by Peter Clarke

Design and layout by Lodestone Publishing Limited and Contentra Technologies Ltd
Illustrated by Rachel Annie Bridgen; www.shootingthelight.com
Cover design by Sarah Duxbury and Paul Oates
Project managed by Sonia Dawkins

FSC
www.fsc.org

MIX
Paper from responsible source
FSC™ C007454

This book is produced from independently certified FSC™ paper to ensure responsible forest management.

For more information visit:
www.harpercollins.co.uk/green

Contents

Numbers to 10

- Trace the numbers. Start at the dots.

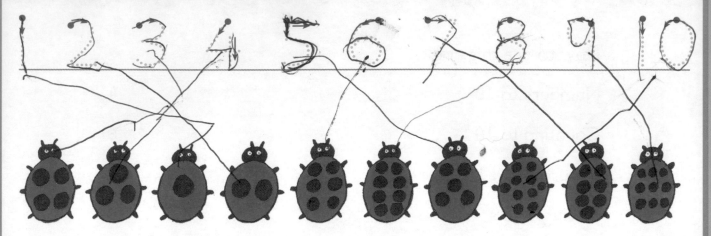

- Join each number to the same number of spots.

Join the dots

- Join the dots in order from 1 to 10.

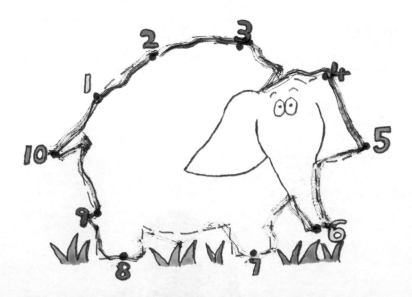

What you need to know At this stage your child is learning to:
- recognise and **write** the **numbers** 1 to 10
- relate a number to a quantity
- know the sequence of numbers 1 to 10.

Game: Cover the spots

You need: paperclip, pencil, 10 counters (or buttons).

- Spin the spinner (see page 2). What number does it stop on?

- Put a counter on the dog with the same number of spots.

- If the dog is already covered, spin again.

- The game ends when all the dogs have a counter. How many counters are there?

Missing numbers

- Write the missing numbers.

 1   2   3   4   5

3   4   5   7

6   7   9

Taking it further   Count things around the home, e.g. spoons, socks, apples. Start with up to 5 objects, then increase to 10.

5

Counting to 10

How many?

- Count the animals in each group.
- Tick (✔) the group that has more birds and frogs.

  [ ] birds

 [✔] birds

  [ ] frogs

 [ ] frogs

Counting

- Draw 6 flowers.

- Draw 10 fish.

What you need to know At this stage your child is learning to:
- **count** a group of objects
- understand that the last **number** counted refers to the number of objects in a group
- compare 2 groups of objects, and say which group has **more** or **less**.

Game: Counting dots

You need: 9 counters (or buttons).

- Cover each square with a counter.
- Remove one counter. How many dots are there in that box?
- Repeat for each box.

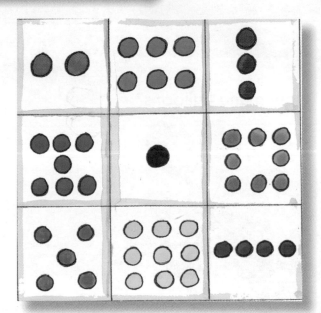

How many bears?

- Colour a plate for each bear.
- Colour a spoon for each bear.

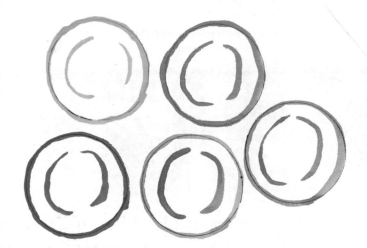

- How many are left over?   plates ☐   spoons ☐

Taking it further Ask your child to **count** objects that: can be touched and moved (e.g. toys); can be touched, but **not moved** (e.g. pictures in a book); cannot be touched **or** moved (e.g. objects far away). Count sounds (e.g. clapping). Count movements (e.g. jumps). Compare two groups of objects. Which has **more/less**?

7

I more and I less

Balloons

- How many balloons? *3*

- How many is I more? *4*

- How many is I less? *2*

- How many balloons? *5*

- How many is I more? *6*

- How many is I less? *4*

More and less

- How many candles? *4*

- Draw I more.
- How many now? *5*

- How many presents? *6*

- How many is I less?  *5*

- Use the number track if you need to.

0	1	2	3	4	5	6	7	8	9	10	11

Game: Tower blocks

You need: 1–6 dice, building blocks or Lego bricks, coin.

- Roll the dice.

- Build a tower with that number of blocks.

- Toss the coin. 'Heads' means add one block. How many now?

- 'Tails' means take away one block. How many now?

- Continue until the tower falls over!

Number puzzle

- Write a different number from 1 to 10 in each ☐.

- Then write 1 less in the ◯, and 1 more in the △.

Adding 2 sets

What is the total?

• Count each set. Add them together.

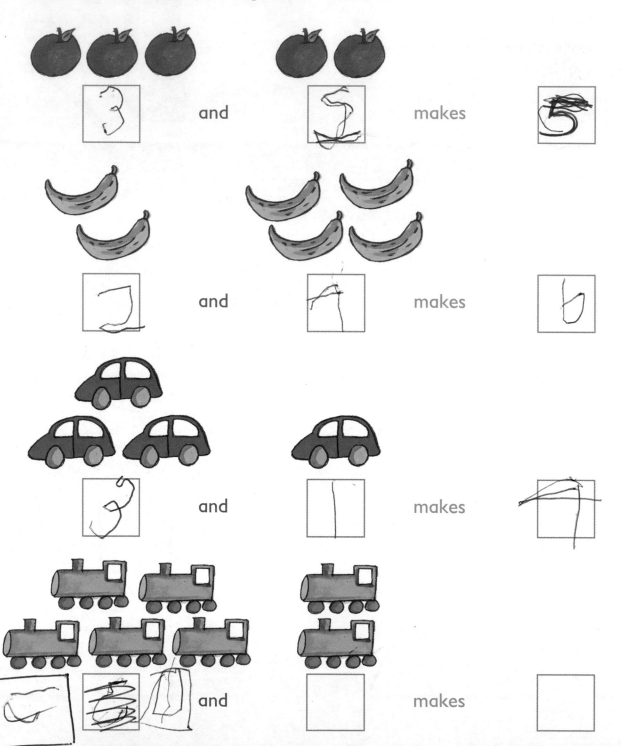

3 and 3 makes 5

7 and 4 makes 6

3 and (blank) makes 4

(scribbles) and (blank) makes (blank)

What you need to know At this stage your child is learning to **count** 2 groups (sets) of objects for totals up to 10. Although the **addition** sign (+) is introduced later, it will help if you use the addition words: **add, total, and, sum, more, makes, altogether**. Ask: 'What does 3 and 4 make?', '**How many** is 3 and 4?', '3 and 4 is how many altogether?'

Game: Making totals

You need: 1–6 dice, 20 counters (or buttons).

- Take turns to roll the dice twice.
- Count the number of dots each time. Add the numbers together.
- The person with the larger total takes a counter.
- Carry on until one person gets 10 counters.

3 and 4 makes 7

Adding the dots

- Roll a 1–6 dice. Draw that many dots in the red box. ☐
- Roll the dice again. Draw that many dots in the blue box. ☐

☐ and ☐ makes ◯

☐ and ☐ makes ◯

☐ and ☐ makes ◯

☐ and ☐ makes ◯

- Count both sets of dots.
- Write the total in the green circle. ◯

Counting on

4 add 2 makes | 6 |

• Use the number track to add these numbers.

5 add 3 makes 8 ✓

2 add 2 makes ✓

1 add 4 makes

2 add 7 makes

6 add 4 makes

What you need to know At this stage your child is learning how to link addition to counting on for totals up to 10. Keep using the addition words (see page 10). When your child is adding 2 numbers together, encourage them to hold the larger number in their head, and to **count on** from that number.

Game: Flower totals

You need: 1-6 dice, 12 counters (or buttons).

- Roll the dice twice. Add the 2 numbers.
- If the total matches one of the flower numbers, cover it with a counter.
- If not, put a counter on the flowerpot.
- If you get the same total, roll the dice again.
- Can you cover all the numbers before you put 6 counters on the pot?

Quick quiz

 makes

 makes

- Count on 2 more than 5.

- What is 3 more than 4?

- What is 3 add 3?

- What is 6 add 1 more?

Taking it further   Give your child more practice in counting on. Point to another number on the swamp number track on page 12, e.g. 6, and ask your child similar questions as in the Quick quiz above.

Taking away

Take away 1

• Cross out 1. How many are left?

5	take away	1	is	4
2	take away	1	is	1
4	take away	1	is	3

Take away 2

• Cross out 2. How many are left?

| 5 | take away | 2 | is | 3 |

| 3 | take away | 2 | is | |

| 6 | take away | 2 | is | 4 |

What you need to know   At this stage your child is learning that **subtraction** involves taking away objects, and counting **how many are left**. Although the subtraction sign (−) is introduced later, it will help if you use words like: **take away, less, difference between** and **leaves**.

Game: Spinner take away

You need: paperclip, pencil, 10 counters (or buttons).

- Put 10 counters on the table. Count them out.

- Spin the spinner (see page 2). What number does it stop on? Remove that number of counters, saying e.g. '10 counters take away 3 counters leaves 7 counters'.

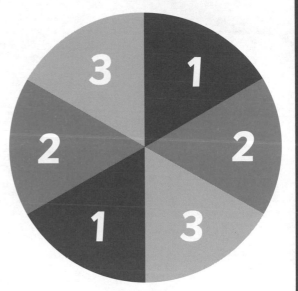

- Carry on until there are no counters left.

How many are left?

7 birds take away **3** birds leaves 4 birds.

6 frogs take away **2** frogs leaves 4 frogs.

Taking it further Count out 5 counters. Hide 2 of them. Ask: '**How many are left?**', '**How many have gone?**' Repeat with other numbers.

Counting back

How many are left?

5 take away **3** leaves **2**

- Use the number track to help.

4 take away **2** leaves

8 take away **5** leaves **3**

7 take away **3** leaves **4**

5 take away **2** leaves 3

6 take away **4** leaves

What you need to know At this stage, your child is learning how to:
- link subtraction to counting back from a larger number in order to find **how many are left**
- take away from a set of less than 10.
Carry on using the subtraction questions (see page 15).

Game: 5 hungry mice

You need: 1-6 dice, 10 counters (or buttons).

- Roll the dice twice.

- Take the smaller number away from the larger number.

- If that number matches one of the mice numbers, cover it with a counter. Then repeat.

- If you get the same answer, put a counter on the cheese.

- Can you cover all the mice before you put 5 counters on the cheese?

Egg puzzle

- There are 6 eggs in a full box.

  How many are left in each box?

 4 eggs    2 eggs

 3 eggs    5 eggs

Taking it further   Point to a number on the number track on page 16, e.g. 7, and ask: '**Count back** 3 from 7', 'What is 3 **less** than 7?', 'What is 7 **take away** 3?' Then encourage your child to ask *you* similar questions.

Coins

- Match the coins.

How much is it?

- Match the coin to the price.

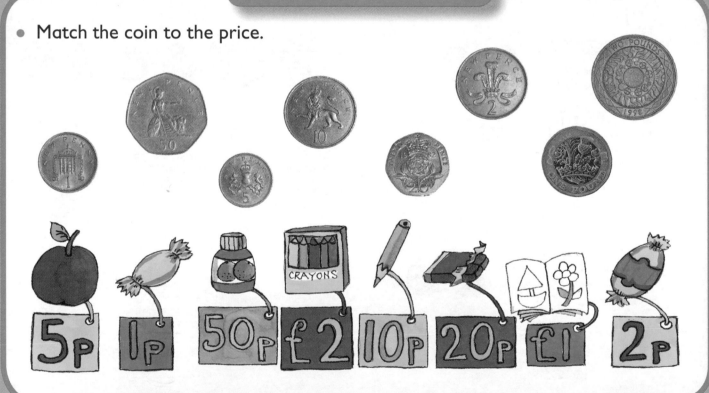

What you need to know At this stage your child is learning:

- to recognise and sort coins, including £1 and £2
- to find **totals** of amounts up to about 10p, and understand the idea of **change**
- that some coins have a greater **value**, and will buy **more**
- to exchange a group of coins for another **coin**, e.g. two 1p coins for a 2p coin.

Game: Saving 10p

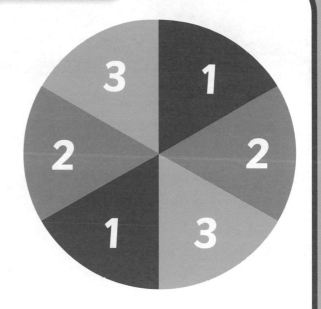

You need: paperclip, pencil, 10 1p coins.

- Spin the spinner (see page 2). If the spinner stops on e.g. 2, cover 2 coins on your moneybox using 2 of the 1p coins.

- Spin the spinner again. If it stops on e.g. 3, this time say:

  '2 pence add 3 more pence is 5 pence'.

- Carry on until you have ten 1p coins in the moneybox.

Making 10p

- These 2 purses each have 10p in them.
- What other ways can you make 10p?

Taking it further Ask your child to **sort** coins, e.g. by colour, shape and **value**. Then take two 1p coins, and ask your child to find a **coin** that is **worth** the same, i.e. a 2p coin. Try other coins, e.g. five 1p coins, and a 5p, etc.

Comparing

- Tick (✔) the shorter scarf.

- Tick (✔) the taller animal.

Heavier or lighter?

- Which is heavier? (✔)

- Which is lighter? (✔)

What you need to know At this stage your child is comparing 2 **lengths**, **widths**, **heights**, **depths**, **weights** and **capacities**.

Game: Let's compare !

You need: paperclip, pencil.

- Spin the spinner (see page 2). When the spinner stops on an object, ask is it longer/shorter; bigger/smaller; fatter/thinner than the one opposite.

- Carry on until you have landed on each picture.

Which is it?

- Who is lighter? (✔)

Mark

Pete

Layla

Mark   Layla

Pete   Dan

Dan

- Which holds more? (✔)

Patterns

- Continue the patterns.

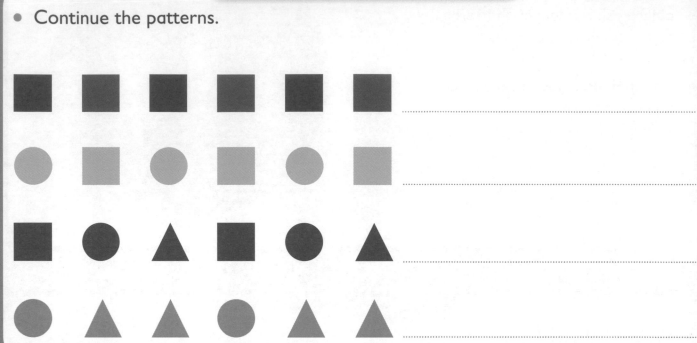

Missing shapes

- Draw the missing shapes.

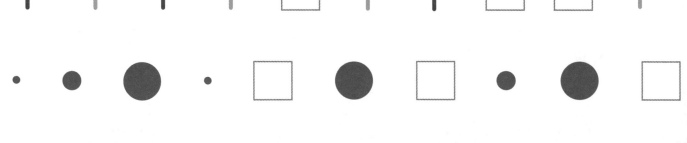

What you need to know At this stage your child should be recognising and creating simple patterns using different sizes, shapes and colours, e.g. a, b, a, b; ○□○□; a, b, c, a, b, c; ✳✳✳✳✳✳; a, b, b, a, b, b; △△△△△△.

Game: Animal line dance

You need: paperclip, pencil, 4 counters (or buttons).

- Spin the spinner (see page 2). If the spinner lands on the next animal in the pattern, put a counter on it.

- Say the names of the animals in sequence each time you add a counter.

- If it's not the right animal for the next one in the pattern, spin the spinner again.

| sheep | goat | horse | cow | sheep | goat | horse | cow | sheep | goat |

Party patterns

- Continue the patterns on the party hats.

Taking it further Make repeating patterns, using e.g. 6 knives, 6 forks and 6 spoons, e.g. knife, fork, knife, fork. Ask: 'What comes next? Why? Can you continue the **pattern**?' Then try with forks and spoons. Encourage your child to make repeating patterns of their own e.g. threading beads on a string.

Solid shapes

- Match the shapes.

Game: How many faces?

You need: 1–6 dice, 8 counters (or buttons).

- Roll the dice.
- Find the shape with the same number of faces (surfaces), and put a counter on it.
- Carry on until all the shapes are covered.

- Colour the spheres yellow.

- Colour the cubes pink.

- Colour the cones orange.

- Colour the cylinders blue.

- How many?

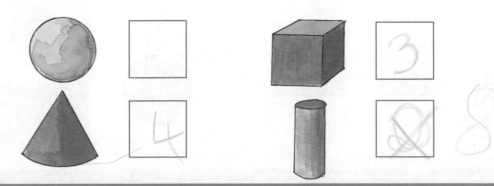

Flat shapes

- Match the shapes.

Game: How many sides?

You need: 1–6 dice, 8 counters (or buttons).

- Roll the dice.
- Find the shape with the same number of sides, and put a counter on it.
- Carry on until all the shapes are covered.

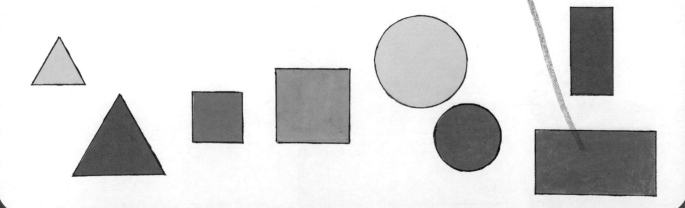

What you need to know At this stage your child is learning to:
- recognise, name and describe simple **flat shapes**, (**circle**, **triangle**, **square** and **rectangle**)
- count the number of **sides** of flat objects
- recognise examples of flat shapes in the world around them.

- Colour the triangles red.
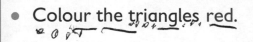

- Colour the rectangles blue.

- Colour the squares green.

- Colour the circles yellow.

- How many?

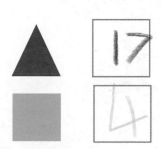

	17		4
	4		

Taking it further Ask your child to find things at home that are shaped like a **circle**, **triangle**, **square** and **rectangle**. Encourage them to draw one of the shapes, then to draw it again **larger/smaller**. Ask them to describe a circle/triangle/square/rectangle, saying how many **sides/corners** each one has.

27

Above and below

- What is under the table? Colour it brown.
- What is on the table? Colour it orange.
- What is above the sink? Colour it blue.
- What is outside? Colour it green.
- What is between the sink and the door? Colour it red.

What you need to know   At this stage your child should be using words such as: **over, under, above, below, on, in, next to** and **beside** to describe where things are. Talk about where things are in the picture using the position words above, plus **left, right, before, after, outside, inside, in front of, behind, between, opposite, around.**

Game: Left or right

You need: a counter (or button), a coin.

- Put the counter on the star, and toss the coin.
- If it is 'heads', move the counter to the left.
- If 'tails', move the counter to the right.
- Carry on until the counter reaches or

On the shelves

- Draw a ☕ above the ▯ .
- Draw a 🫖 to the left of the ☕ .
- Draw a 🥣 below the 🫖 .
- Draw a ☕ between the 🫖 and the ☕ .
- Draw a 🫖 to the right of the ▯ .

Taking it further Play I spy … something **above** the television/hanging **on** the wall.

29

Sorting and matching

Where do you find them?

- Match the objects to the rooms.

Matching pairs

- Match the socks.

What you need to know At this stage your child should be sorting, matching and ordering things. Start by sorting into round and not round, square and other shapes, then **sort** using their own suggestions.

30

Game: Sorting by colour

You need: paperclip, pencil.

- Spin the spinner (see page 2). What did it stop on?

- Find that number of small items in that colour from around the home.

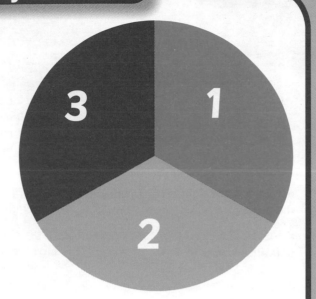

What belongs to which?

- Match the boxes to the lids.

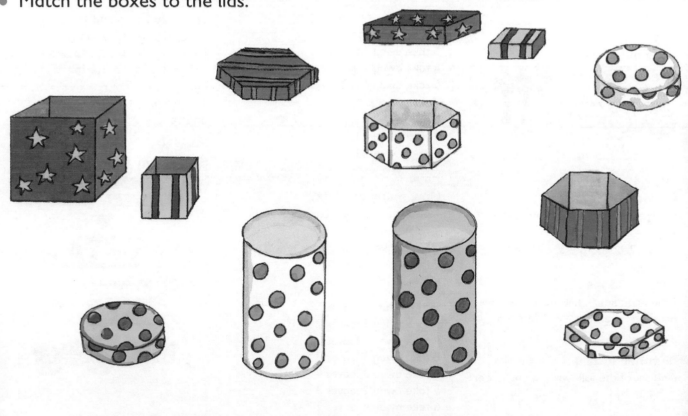

Taking it further Find opportunities around the home for your child to **sort** or **match** objects, e.g. clothes, buttons, toys, books… Suggest they sort by shape, size, colour, or how and when they would be used.

Answers

Page 4
One to ten
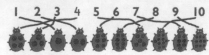

Join the dots
Check that your child has joined the numbers in the right order to complete the picture.

Page 5
Missing numbers
1, 2, 3, 4, 5
3, 4, 5, 6, 7
6, 7, 8, 9, 10

Page 6
How many?

Counting
Check that your child has drawn 6 flowers and 10 fish.

Page 7
How many bears?

plates = 2, spoons = 1

Page 8
Balloons
Picture 1: balloons = 3, 1 more = 4, 1 less = 2
Picture 2: balloons = 5, 1 more = 6, 1 less = 4

More and less
Cake: candles = 4, 1 more = 5
Presents: presents = 6, 1 less = 5

Page 9
Number puzzle
Any six of the following in any order:

Page 10
What is the total?
3 and 2 makes 5
2 and 4 makes 6
3 and 1 makes 4
5 and 2 makes 7

Page 11
Adding the dots
Check your child's answers.

Page 12
How many altogether?
4 and 2 makes 6
5 and 3 makes 8
2 and 2 makes 4
1 and 4 makes 5
2 and 7 makes 9
6 and 4 makes 10

Page 13
Quick quiz
1 and 4 makes 5
2 and 3 makes 5
2 more than 5 = 7, 3 add 3 = 6
3 more than 4 = 7, 6 add 1 more = 7

Page 14
Take away 1
5 take away 1 is 4
2 take away 1 is 1
4 take away 1 is 3
Take away 2
5 take away 2 is 3
3 take away 2 is 1
6 take away 2 is 4

Page 15
How many are left?
7 birds take away 3 birds leaves 4 birds.
6 frogs take away 2 frogs leaves 4 frogs.

Page 16
How many are left?
5 take away 3 leaves 2
4 take away 2 leaves 2
8 take away 5 leaves 3
7 take away 3 leaves 4
5 take away 2 leaves 3
6 take away 4 leaves 2

Page 17
Egg puzzle
4 eggs, 2 eggs
3 eggs, 5 eggs

Page 18
Matching

How much is it?

Page 19
Making 10p
There are 11 different ways:
10p
5p and 5p
5p, 2p, 2p and 1p
5p, 2p, 1p, 1p and 1p
5p, 1p, 1p, 1p, 1p and 1p
2p, 2p, 2p, 2p and 2p
2p, 2p, 2p 2p, 1p and 1p
2p, 2p, 2p, 1p, 1p, 1p and 1p
2p, 2p, 1p, 1p, 1p, 1p, 1p and 1p
2p, 1p, 1p, 1p, 1p, 1p, 1p, 1p and 1p
1p, 1p, 1p, 1p, 1p, 1p, 1p, 1p, 1p and 1p

Page 20
Shorter or taller?

Heavier or lighter?

Page 21
Which is it?
Layla, Dan